AF443749

The Story of Lickel Bird

A bird's eye view into the private life of a robin

The Story of Lickel Bird

A Bird's Eye View into the private life of a robin

Catherine Bootland

http://www.fast-print.net/bookshop

THE STORY OF LICKEL BIRD
Copyright © Catherine Bootland 2017

All rights reserved

The right of Catherine Bootland to be identified as the author of this work has
been asserted by her in accordance with the Copyright, Designs and
Patents Act 1988 and any subsequent amendments thereto.

A catalogue record for this book is available from the British Library

ISBN 978-178456-463-6

First Published 2017 by
Fast-Print Publishing
of Peterborough, England.

Introduction

I want this story to sound real because it is. It is the true story of a very unusual little robin with an endearing and confiding nature, who has built his life in my garden among the old fences, hedges, shrubs and trees, stacks of pots and bricks and the vine on my balcony, and who took a chance to get close enough to a human for us to build a relationship. Nothing is over sentimentalised and it is told just as it happened over several months. There were other robins around but Lickel Bird or LB as he came to be known, responded to human company and showed remarkable trust and attachment. He watched me in my garden long before he trusted me enough to see I was on the side of small birds and that I would be there for him and what he needed most in his busy life was food.

During my daily contact with him on his countless visits to me, I have observed him performing his many and varied tasks. These observational snapshots of his behaviour during these visits over the spring and summer have

given me a window into his world and have allowed me
to share some of his busy life in all its ups and downs.

I believe I can fill in the gaps behind the scenes and shed
light on what he has been doing and how hard he has had
to work to keep himself and his family alive. I have seen
him toil tirelessly throughout. How he has coped with
finding and maintaining the right nest site, building
it strong enough with his mate to survive all weathers &
support her and the eggs as they hatched and grew into
fledglings remains a mystery, as I have not dared to violate
his secrecy or trust by searching the hedgerow for signs
of him and his little family. All I have to go on are my daily
observations of his routines and activities in my garden,
my encounters with him during his many visits to my
balcony and the moods he has displayed during these
brief but mesmerising moments. LB has a unique
personality, he can be cautious, nervy, thoughtful,
fierce, playful and very engaging and my days are
the richer for being able to get to know him. I would
like to think it was the same for him.

the lives of robins are so short and they work so hard to keep themselves alive, raise their families and pass on their survival wisdoms to the next generation. I'd like to think a memory of me has been passed on too and that one day a descendant of LB will venture onto the curvy vine and look in through my window. observing how fragile their fight for survival is and how alert the robin needs to be to learn about and overcome the constant dangers and anxieties he faces is very humbling. Robins show such devotion, courage and cleverness, how can we not respect and give such hard working and social little birds as sympathetic an environment as possible so that they can live their lives and be able to raise the next generation. My garden would be so lonely without this little red breasted songbird and his willingness to share his life.

the story of Lickel Bird

A bird's eye view into the private life of a robin

Lickel Bird or LB as he came to be known is a bright-eyed inquisitive little robin who lives in and around my garden.

I first noticed him last summer while I was watering my plants in the heat wave. He had seen me before I became aware that he was close by perched on a plant stalk watching the steady stream of water arching from the hose as I directed the water to the base of the plant. He was studying the newly wet patches of ground to see if the wetness would liven up a grub or make a worm wriggle so that he could snatch it. LB seemed to be mesmerised by the glinting water jet as I moved the flow along. He was trusting me enough to follow close to the arching water stream and hoping it would travel in a safe trajectory and not suddenly stop and shower him with water. I gently moved the water jet along so that he could follow and hoped that he would be rewarded for the risk he was taking. His eye spotted something tasty on the soil not visible to me and he darted forward to pick up the morsel in his beak. He did not fly away and continued close by to watch further and wait to see if anything else would turn up.

My digging over some ground brought him very close and he eagerly watched the newly turned lumps of soil for signs of life. Large worms frightened him and he took several steps back when I placed one wriggling at his feet but he became very alert and excited when he saw a centipede. You have to be very fast to catch a centipede before it burrows back under the earth so I picked one up quickly and placed it close by his little feet. He stared at it for a second and knowing he had to be quick, picked it up in his beak and pinched it several times to disable it and stop it making a bid for freedom. When he was satisfied it had stopped wriggling he swallowed it and looked very pleased with himself. LB now realised that I knew what he liked to eat and could get it for him.

I have constructed a bird feeder to support all the little birds in and around the garden and top it up daily with various seeds. This constant food supply is eagerly awaited by a selection of garden birds and is particularly popular in cold weather. LB watched me fill the feeder from his vantage point on top of a garden cane so he knew there was a supply of bird food in a container at the top of a pole. But robins don't care to hang on feeders and peck at seeds like the tits and parrots, they prefer privacy at meal times. As an extra offering, I filled the plastic lid of the container with seed to scatter on the ground under the hanging feeder to nourish any birds who might have vertigo and be unable to reach inside the feeder hanging aloft. One morning after I had scattered the extra seed from the lid which I was holding, LB flew over and hopped onto it and hitched a ride as I carried him perching on it back to my balcony table where he had seen the top-up bags of seed. When we got there he hopped onto the plastic fence which ran round the edge of the balcony and watched as I dipped my hand into the seed bag and pulled out a bird 'nibble', advertised as rich in suet and tiny flies, energy food which birds find irresisible. He bent his head towards my fingers which were holding the nibble up to him and he took it in his beak and swallowed it in front of me. I watched it go down as the front of his red plumage rippled with the bulge. I could see he enjoyed it and wanted more. I offered him another one and then another and then he

paused as the nibbles went down. He was full after three
so he turned round and launched himself off the fence
and was gone. I hoped an addiction to nibbles would br
ing him back so we could get to know one another.

And so it was that LB began his numerous visits to my
balcony. I believe he came to look for me because I cou
ld offer instant food on cold days to ward off his hunger.
I wanted him to become trusting and tame enough to
come regularly for food so that we could spend a few
moments with each other every day. I hoped that as his
visits increased he would see me as his ally, another
living creature in the natural world that we both
inhabited. And as his life progressed, these routine vis
its became a way of life for us both.

Last year I had twisted a frond on the vine growing in
front of my window to make a circle shape which
had hardened up over the winter. LB found that this
curved bit of vine provided the perfect perch and he

loved to land on it. It gave him a bird's eye view up to my
balcony window while the surrounding vine leaves gave
him added cover and protection as he waited to be fed.

This curvy branch of the vine became his domain and
constant perching spot when he flew down from the nea
rby yew tree branches. It was the perfect size for his thin
spindly feet to grip and also gave him a view in through
the window which he realised was the opening to my
world. From just inside the room I could watch him

land on the curvy vine and stretch up to peer in to see if I
was there. He knew as soon as I saw him I would emerge to
offer him nibbles from the small, shiny bowl I kept them
in. Sometimes he was so eager for the food, he flew straight
onto grip the side of the bowl while I was still emerging
from the doorway and would land on my finger & bend
his head down to view the little pile of nibbles before
quickly selecting one for himself. If he landed on the
edge of the bowl, I always let it be self-service. This way
of taking food was only one of his many little routines
which he adopted depending on his mood and where I
would be sitting or standing when he dropped by.

His visits to the curvy vine became more and more freq
uent. As soon as he landed, I showed him the nibble bowl
and selected one to hold up to his beak for him to take
so that he would always have an expectation of receivi
ng food when he came to the balcony. A bribe to convin
ce him that coming to see me was worth his while.
He watched very attentively and I could see that he
was nervy and hoped that the expectation of food

would enable him to overcome his fear. LB would swall
ow one and fly away with the second and very soon he
developed a nibble addiction as I hoped he would, land
ing on the curvy vine branch at various intervals th
roughout the day, pausing to stare at me before bend
ing his head forwards to get hold of the nibble I held
up to him. He used quite a lot of force in the beginning
as though he expected me to hold on to the treat but
gradually he took the nibble carefully and with just
the right amount of pull. He swallowed it quickly and
waited very still looking at me as if transfixed for a
few moments while the nibble went down his throat
before he was ready to bend over for the next one. This
was a much bigger bite than a small squishy insect
egg and his throat would not be very wide so he wou
ld feel its impact and need to wait for it to settle.

The routine of daily nibble swallowing extended into wee
ks and during those times when LB flew down & landed
on the curvy vine hoping and expecting to be fed we have
bonded. He has learned to watch for the times when he

knows I am at home behind the window. Throughout the
cold winter months, a roller blind covered the balcony win
dow excluding the view out and the view in. LB would sit
in the branches of the tall yew tree in sight of the covered
window and wait for me to roll up the blind and draw
the curtain. These movements and the sound they made

let him know I was ready to open up the door & emerge.
As I rolled up the blind and looked out, I could see him bob
bing up and down in anticipation of the door opening
and my emerging with the nibble bowl. He waited for
the sound of the door latch to open and then immediat
ely flew down onto the curvy vine branch looking up to

meet me and eager to be in position ready to bend his head forward to grasp the nibble with his beak and swallow it, pausing to wait before bending again to take another. I broke them into small bite sized portions so he could eat several which would take a bit longer and would allow me time to talk to him and get him used to my voice and movements. The cold made him hungry.

I decided to announce my appearance on the balcony by making a distinctive vocal call so that LB would learn to associate the call with food and my presence and hopefully learn that this was a call to alert him to come if he wasn't already in the vicinity. I had seen LB fly all around and beyond the garden so knew he had a wide territory that he explored. Robins sing distinctive songs high up in trees to be heard far and wide so I also knew that LB had very good hearing. I wanted him to associate my call with food availability and be able to deduce that when he heard the call he would know I could be seen and that it was safe to fly over and land on the curvy vine branch. I felt

a big responsibility for keeping him safe and scanned arou
nd the garden for any unpleasant creatures that could ha
rm him before calling. But LB was very cautious and hig
hly aware of his surroundings and knew when it was sa
fe to descend. I repeated his name Lickel Bird again and
again in a singsong tone followed by a clicking sound wh
ich I hoped was attractive to robins. This amused me to do
and I perfected a mellifluous tone which was very disti
nctive and I could see that he listened to it. He learnt
quickly that this musical call proclaimed my presen
ce and he responded. I always use this call to alert him
and he usually comes within a few minutes if not imm
ediately. This is very useful if I have been out for the day
and LB has only seen drawn curtains and has given up.
On these occasions my call brings him quickly to the
curvy vine branch eager to be given some food.

LB has his own particular routines and he was starting
to understand mine. I like to think my call is my way of
talking to him when he is perching next to me. It seems
to calm him down, he is always a little nervous when he

flies down and he stares at me when I say the singsong
words so I think he likes it but it maybe he just tolerates
it while waiting patiently for food. On a few occasions
he has sung back with a throaty wheezy chirp but he
is always on the lookout for threats and doesn't tarry
so his song is very short. Sometimes LB has a particu
lar song still in his throat that he continues with af
ter he has landed on the curvy vine and I wonder w
hether it is a left over warning song delivered to a ri
val or whether it is a friendly song intended to beguile.
If LB had been engaged in a territory dispute before he
landed, his song would act as a warning to his rival.
His spring song is very powerful and can be used to
defend his home and also to attract a mate to share it
with him. I wanted to believe that this warble-wheezy
song was just for me & that he felt safe enough to perform
it.

Feeling safe cannot be guaranteed in my garden. The
re are many cats who do not belong to me but to neig
bours who put them out all day to roam and stalk ar
ound in other people's gardens. They are not welcome

in my garden which is a bird sanctuary. I have watch
ed them sneak in under the fence and down tree bra
nches and from other vantage points always peering
around for illicit action. They trade on the fear they
instil into small creatures and their constant prow
ling for birds to ensnare means that small birds must
be constantly on the lookout. These discerning little he
roes have learned to watch for the sign of the cat and
know what to do to avoid disaster.

LB knows about cat danger and how to scan the horizon
for the enemy. When he senses danger he puts his back
side in the air cocking up his tail feathers and scruti
nises the shape of the danger he has fixed his eyes upon.
He decides what to do in a second, sometimes he waits very
still for the danger to pass by with eyes focussed, & som
etimes he flies off. The danger can be a cat, a fox, a big
bird or even another robin. LB comes from an adjacent
garden and I have seen a resident robin in my garden
who is very different in his behaviour. Territorial frict
ion can quickly erupt. This other robin has occasionally

been in view when LB has landed on my fence for fo
od. They spot each other and this recognition unnerves
LB who stiffens up to face the threat. He starts to sing
a warning song, loud and direct and has even sung
this warning song to the other robin while holding
a nibble in his beak. His throat can cope with both
actions simultaneously, swallowing and singing. He
seems very brave and courageous to me at these
times, being only a few centimetres high.

LB is not afraid of eye contact. I wonder what he is thi
nking when we stare at each other as I say his name,
Littel Bird over and over in as many soothing ways I
can invent to make him feel safe. When he is looking in
to my face with his beady black eyes I think he is trying
to puzzle me out. Am I a much larger version of him
in my red jacket which I wear to encourage that thou
ght. Is he wondering why my eyes are so much bigger
and why I hold up food to him? But LB doesn't deliber
ate for long, he is always flying off somewhere, alwa
ys on a mission. I sometimes think that being in mis

sion mode will wear him out and he will expire from exhaustion. The male has to feed the female in courtsh ip to secure her affections and then continue feeding the hatched brood if courtship ends with successful progeny. LB must keep going and take on all these res ponsibilities which are his destiny. He instinctively m aintains his routines and there seems little time to pause. I hope he has time to be and time for me too even though his visits to me last only a short amoun t of time each day. He can't know that his life may not last very long. The average lifespan for a robin is 1·7 years and for LB it is a statistic I want to demolish ε the only way I can do that is to feed him up and keep him well nourished.

Around this time he swallowed one nibble and if he was very hungry he swallowed two. Then he took another wh ich he didn't swallow but carried in his beak to a yew tree branch and tapped it several times. I could hear the tap of the nibble as he struck it against the branch. Was he testing it in some way to make sure it was acceptable as

I imagine that this bite was destined to be popped into the beak of his mate. She is needy and obviously picky and he wants to please her. I have seen them together on the yew branch just above my balcony, she flutters close to him and begs for morsels. LB collects nibbles from me and prepares one that passes the standard and takes it to her on the nearby branch where she is waiting. He deli cately holds the nibble, goes up close and pops it into her beak. She is pleased with the gift and swallows it but she is very shy compared to LB so doesn't stay long and quick ly flies off presumably back to the nest. I don't know where their nest home is, out of sight somewhere secret and well hidden I hope. I feel touched that he has brou ght her to meet me and to share the food.

One time when I was sitting on a balcony chair staring into the garden, LB suddenly landed close by on the pla stic fence and his little spindly feet made a resound ing plonk as they hit the plastic. He liked to surprise me and was carrying a small green grub in his beak. He stared into my face and would not swallow the grub which he needed to do before he could accept the torpedo shapped nibble I was holding up to him. Was he offering me the green grub in exchange for the nibble, his way of rewarding me and acknowledging that I was one of his family? I'd like to think this was the case but a later episode has shed light on this behaviour. Well into spring when insects and grubs are more abundant LB landed on the fence carrying a squashed spider in his beak with its legs sticking out. As he opened his beak to cram in a nibble the crushed spider dropped out. He hadn't realised that in order to open his beak wide enough to secure the second meal he had lost the first but he carried on and flew off to deliver his haul into a hungry mouth waiting in the nest.

LB doesn't like wind or rain. He knows that rain will make his feathers damp and the motion of strong winds will distort the levels of the branches that he flies between to carry out his routines. One particular morning the wi nd was very strong causing the branches to sway but then eased off allowing them to be still again.whencoming for food LB usually bides his time in the branches of the yew tree above the balcony, waiting for the right time to drop down onto the curvy vine to stare up at the window to alert me that he is there ready to receive. But as time went by he didn't appear which was unusual as LB liked his routines. The hours went by and I began to think that he had perished. Roaming cats are persistent prowlers in my garden and intent on mauling anything that they can snare and are a menace to small birds trying to live their own private lives. This violent scenario took hold of my thoughts and I began to grieve that I would never see LB again. I went down the garden calling him but he wasn't there. I sat in my balcony reading late into the afternoon feeling the loss of my little friend w hen suddenly he landed on the vine and flew over. I was so overjoyed that my greeting must have sounded very excitable. Had hunger forced him out of his hiding place to risk flying in bad weather? He wouldn't know I was worried about him. A cobweb was clinging to his dry red breast feathers so he must have been sheltering somew here secluded. LB knew when to lie low out of danger and stay hidden, this was a trait I would recall many times

when I became worried over his absence. He was careful, cle
ver and resourceful and knew how to survive.

Early on cold mornings before the spring sunshine had
started to fatten up the buds, LB was already perching on
the curvy vine waiting for me to open the curtains.
Sometimes when I opened the door he would alight on my
fingers as I carried out the nibble bowl to grasp a nib
ble faster, he was so eager for food. On these occasions, LB
needed to perform his routines at speed and he didn't
have a moment to lose. One very early morning I wan

ted to make an early start on tidying the garden so I was out on the balcony just as dawn was breaking and I had not spotted him waiting in the dim light. He sensed I hadn't seen him so to get my attention he fluttered against my arm which was behind me with his dangling feet to let me know he was there waiting for food. This wafting sensation on my bare arm gave me a fright and I pulled it away quickly not realising that LB was trying to communicate. I hoped my swift arm withd rawal hadn't upset him and immediately tried to rea ssure him with my singsong call. He quickly fluttered around, adjusted his position and hopped onto the curvy vine and swallowed the nibbles I held up for him.

In the early days of spring when the sunshine beamed down and there was no wind, LB would bring along his little red breasted mate who joined him in the yew branches close by for a gourmet double act. She hove red fluttering her wings and fluffing out her feathers. She was not as bold as LB and needed to be coaxed and fed beak to beak. LB is eager to please her and flies from spot to spot searching out creepy crawly morsels to take back to her as she waits on the branch where she looks rather fragile and a bit of a drama queen. He usually offers her nibbles collected from me as well which he takes up to her and tiptoes close enough to pop them in to her beak. But she is a bit careless and too choosy and she often lets go of his offering. In a flash, he follows the trail of the dropped nibble and dives down to the ground to find its landing place. He picks it up and flies back to her and reoffers it but I detect signs of frustration with her as he sometimes swallows it himself. But little mate flutters her feathers and hovers closer to be given another chance.

One sunny evening when I was expecting LB to land on the vine, I saw him hopping on the branch close to his fluttering little mate who was demanding all his attention and to be fed. LB was striking at evening midges too small for me to see but scoring & putting his catch straight into little mate's beak. She hopped ever closer expecting more bounty and LB knew that it was his job to supply. It seemed as though he was trying

to show her what to do so that she could learn to do some of
the catching while the midges were plentiful. The sun lit
up the crevices in the bricks and the couple hovered from
crack to crack where the midges were hiding. But little
mate held back, it was obvious she didn't want to do any
of the catching, she just wanted to be pampered and for a
ready meal to be dropped into her beak. This could be her
ultimate mate acceptance test, would LB be prepared to
keep her nourished, content and constantly supplied. I
think he must have passed the test as they are still tog
ether. All he needs to do now is to prove he can be her fait
hful and exclusive consort for the immediate future,
robins don't pair for life, they don't know what lies ahead.

All his skills would be needed now and his shy fluttering
mate would be depending on him for several weeks. His
daily routine had changed so they would have completed
their nest by now which would have to be done before she
felt ready to ovulate and develop her eggs. She must be
well nourished for all to go smoothly. As she helped to
build the nest she will have memorised all the features

around so that the whole nest site environment would be in her visual memory. As she sits on the eggs, LB will feed her constantly so that she can maintain sufficient warmth to incubate them and ward off the chill if the nights are cold. Maintaining the right temperature for the eggs to develop is vital. LB needs to keep himself nourished too. Both birds will have invested a lot of time and energy into building their nest and filling it with eggs. LB must work very hard now for the clock is ticking. He must find all the food, guard and defend the nest site and fit in his daily nibble collecting from me. I want to think he likes to see me when he drops in, he usually pauses for a quick look at me before bending down to take one or two nibbles into his beak to swallow. But he doesn't linger, he flies off with one in his beak for his little mate, so I know that she is alive and well and the nesting is on track.

May is here now and foliage everywhere is almost bursting out as you watch. In amongst the new leaves there should be freshly hatched caterpillars and grubs for LB to find. I suspect the cold mornings have thwarted availability of this grub glut and LB has grown accustomed to nibbles. I hope the addiction is strong enough to keep him coming to the curvy vine. If and when he has hungry hatchlings to nurture, he will be foraging all over the garden and perhaps the pressure to provide as varied a diet as poss ible will force him to take risks. There are so many spie s in the garden not least the creeping cats and thieving magpies. But I know LB is clever and his beady eyes spot danger before I do, he was a fast learning fledgling.

I wanted to take a picture of LB on my digi camera to have with me when I was away from the garden. After the night-time he is hungry and hopefully will be able to wait a little longer for food which would give me a cha nce to take a photo of him. I knew he would be perching on the curvy vine first thing in the dawn light waiting for me to open the curtains. Holding up a camera which is bigger than a robin might be scary for him but I held it up in one hand and held the nibble for him to take with the other. This little scene would be my shot & some initial success made me want to hold the camera long

enough to make a little movie of him. The next fine mo
rning when LB landed, I tried to get the camera in position
but he got tired of waiting and hopped onto it and gripp
ed it with his little feet round the frame looking perplexed.

I had to reward him immediately with a nibble so that
he wouldn't lose interest. I needed him to feel secure when
coming for food. Next time I had practised holding a nib
ble in one hand and the camera in the other and my
synchronisation was much better. I now have LB The Movie.

Since my rapport with LB has grown I have become very protective of little birds. As I look around my informal garden, unkind onlookers might say a bit wild, I am always weighing the dangers from his perspective. The more I see the more I worry. There are many small bird attractions, stacks of plant pots for insects to hide in, low level canes propping up plants to perch on, showers of seeds disturbed out of the feeder by clumsy parrots, an uneven lawn full of grub holes, all bird havens eyed up by mean, stealthy cats who creep into the garden to spy & wait. My discouraging tactics work when I see the prowlers but I can't watch over the terrain at all times. There are magpies and crows scanning the ground from high branches waiting to swoop down on unsuspecting birds who are busy searching for their next meal unaware of these marauders. I want LB to be well nourished so that he is sharp and in constant alert mode to be able to spot danger ahead of calamity. I want him to be strong enough to cope with raising the next generation and not be worn out doing it. Remembering a poor robin survival statistic strengthens my resolve to feed him up to help him survive another winter. I wish he could und erstand my concerns so that he would know I am more than just a food supply and that we are both working to survive on the planet, sharing the same days & nights.

I had to leave LB to go on holiday for a fortnight and was worried that he would wait endlessly for me to appear with the extra food he had come to rely on. He was waiting on the curvy vine perch the morning of my departure and he took two nibbles as usual, one to eat & one to take away. Then I drew the curtains and left knowing that the next time he returned for food he would wait and wait in the yew branches ready to drop down onto the curvy vine. He knew that closed curtains meant no one home but he had learned that they would eventually be opened and I would be there with food. But not this time. I was travelling hundreds of miles north passing thousands of other robins but not mine. Each day I was away I knew LB would wait in vain on the vine branch looking up at the covered window & that eventually he would give up on me. I knew he was resourceful and that he would look elsewhere for food.

His survival instinct would be much stronger than his food friendship with me and I would cease to matter in his world. The garden was starting to burst open with new life, green leaves swelling to provide shelter for flies and food for grubs. These ready meals would help him to forget me as day followed day and the curtains remained closed. I started to regret going on holiday. What would I find after many days away, would his routine have changed & would he still remember me?

The first thing I did when I got back was to open the curtains and go out onto the balcony and call for LB, exaggerating my singsong call over and over hoping he would hear me and the call would jog his memory. Robins are sedentary birds living their lives in the same bit of territory they defend as their own so I knew he wouldn't have relocated. The garden was very overgrown and grubs would be fattening up on the abundant vegetation. LB would be enjoying a new menu without nibbles and would have forgotten me and I would be redundant. It was too much to expect that a small wild garden bird would remember

a call that he had not heard for a long time. I called yet
again to proclaim that I was here but still there was no
sign of him. I removed the new growth of vine branches
which were obstructing the curvy branch & which had been
LB's favourite perch. Now once again it was a visible land
ing pad. I called again and then again an hour later. And
then suddenly he appeared but not as directly as before.
My relief poured into my singsong welcome which I re
peated over and over to help him recall his memory of
me. But he was nervous and not entirely sure who I was
or whether he could trust me. He looked away and then
straight at me and then he saw I was holding a nibble,
this was familiar. He hopped closer to my fingers and
bent his head to take it, turned quickly and flew away.
But I could see he knew the drill he knew the curvy vine
he hadn't forgotten, it was him. I was so relieved.

He visited the curvy vine once more that day, I think he
was testing the situation to see if I really was the constant
food friend from before. The garden was busy and noisy.
There was the raucous cawing of magpies in the tree tops,

unwelcome tenants sharing the same garden space. LB would have to be very vigilant. He always spotted a cat if one ventured close by turning his head towards it and staring with his black piercing eyes, weighing up whether to fly off or remain still until the danger had passed.

I don't want large predatory creatures in my garden, they make me feel even more protective of the little birds that make my garden their home and congregate round my bird feeder. LB had never been greedy and always cautious.

He always looked around him checking for dangers before coming close to take nibbles from my fingers. I was growing more confident that he would resume his daily visits & my thoughts lingered on inventing ways to deter unwelc ome creatures that could harm him, from prowling around.

In the days that followed I could see that LB's routine had cha nged, he was very busy and purposeful and I determined that he and his mate were now caring for hatchlings. Feeding them was LB's responsibility, so many sorties here & there had to be fitted into his garden time. There wasn't a spare moment in his day. He resumed his regular drop in visits to the curvyvine and onto the plastic fence to join me on the balcony, plonking down his thin plasticky feet as he landed. He grew familiar again with our routine, bend ing his head to grab the nibble from my fingers, pausing to stare at me briefly as it was swallowed, and then one to take away. As LB gathered a balanced diet for his mate and the hatchlings, he often arrived with a fly or a small grub already in his beak and when he tried to cram in a nibble along with the squashed & wriggling delicacies they would often drop out to make way for the bigger nibble, a shame as some of his catches must have been hardwon.

As the pace of his sorties increased, it became clear that LB was regularly harvesting a selection of creepy crawlies for his newly hatched offspring. I don't know exactly where his nest is I never will, but I know the direction of it as I see him fly fast to the dark dense bit of hedgerow that conceals it. LB flies solo now, his shy mate doesn't accompany him as she did in their courtship days. She will be busy nest brooding and expect LB to forage relentlessly and return with constant meals. When the eggs become tiny helpless hatchlings, she too will feed them regurgitated worms and insects, spending every daylight hour searching for them, returning to the nest every few minutes. It is highly likely that part of LB's hatchling meals are nibbles so they will be well nourished. I sense that the survival of the next generation weighs heavy on LB's daily endeavours as his flying around seems more frantic. I hope he can find respite once the current chick rearing phase is over but it is quite possible that the pair will try to rear

another brood if the weather stays mild as robins are pro
lific breeders. While on the curvy vine, LB has occasionally
warbled a few short song notes before swallowing the
nibbles and sometimes he has warbled a high pitched
throaty note as he flies off. Being so close to his melodi
ous warbling is mesmerising. I am not sure what the
song means, whether it signifies contentment or wh
ether it is the continuation of a dialogue begun before
he flew in. I like to think it is his way of communicating
with me. I always talk softly to him when he warbles,
he seems to like it and watches my face for a few sec
onds before taking the nibbles and flying off.

very heavy wind and rain curtail LB's sorties for live food.
He can't afford to get his feathers too wet and insects hide
when it's cold. He knows I have food and where I am when
he is hungry. One time when I was working inside but in
sight of the window that looked out onto his curvy vine
perch, he flew up to the glass to peer in to see where I was. I
could see his thin little legs hanging down immobile as
his beating wings held him absolutely still like a humming

bird so that he could stare in long enough to get my attention.
This is quite a manoeuvre for a little garden bird, he must
have been very hungry to take such a risk. I took the nib
bles over to where he was balancing in the air and he
steered himself towards my fingers and took a nibble
while still fluttering in the air. What a brilliant trick,
being able to hover so precisely in the air. I hoped his curi
osity would lead him to venture through my doorway again.

LB continues with his routines in light wind and rain. He
has appeared suddenly with damp feathers sticking up
around his head and on his breast, plonking his spindly
feet down on the plastic fence. He can judge the sway pow
er of the wind and stay upright on the fence while it blows
into his soft white under feathers and dishevels them Even
when the wind is still he can appear dishevelled with tufts
of feathers sticking up round his beak and bits of cobweb
clinging to his breast feathers. Sometimes he is carrying
grub trophies in his beak, still wriggling and waiting to
be swallowed. I wonder when he appears with these for
aging spoils whether he is showing me that he is a good

food provider and doesn't need to depend on my offerings.
At other times I wonder if he wants me to sample his
catch in exchange for what I give him. I want to believe

that he can think these thoughts & recognise me as more
than a moving bird table. I want him to think that I am
part of his family and I don't want his memory of me
to fade. The theory of natural selection informs me that
generations of his genes have honed him as a smart
survivor & visiting me is another survival technique.

when coming for food LB never lets his guard down. His eyes
are always watchful and take in the panorama. He pauses
as he perches weighing the moment before he bends his
head towards my fingers to grip the nibble. Sometimes he
takes the food from my fingers, hovering without perching
first and I wonder if he thinks I am too slow or he is just
ravenous. Before the brood came along, he was freer with
his visits taking time to swallow one bite before bending
for the second. I watched the first bite go down and ripple
his feathers as it travelled down his throat. Sometimes he
made a little songlike noise but now there was such an
urgency in his movements and he had cut out these inti
mate little gestures of recognition, he was eager to be off
zig-zagging through the branches on his next mission.

It wasn't long before LB came to peer in through the window
again to get my attention. Many morning vigils perched
on the curvy vine waiting for the curtains to be drawn,
had taught LB that it was only a matter of time before
I would appear at the window. One cold morning in May
before the sun had warmed up the garden grubs, LB was

already waiting on his branch for the curtains to open with a pleading look. I wondered what time he had started his vigil to be ready to catch my attention when I appeared. I automatically looked for him now and quickly gave him a nibble to satisfy his hunger with my singsong greeting. This time after his snack he returned quickly but I had moved and he couldn't see me. LB knew that I must be som ewhere and unable to wait any longer he came right up to the window and peered in through the glass, by which time I saw him pause in mid-flight, frantically searching around with his eyes to get my attention so I would come out to give him a nibble, which I promptly did. I have seen him do this humming bird act several times now. How can such tenacity not be rewarded?

LB had to ensure the survival of his family including his shy quivering mate now brooding their chicks, who only a few weeks ago was being fed treats beak to beak in the full flush of courtship. He darted back and forth like a torpedo to the nest taking nibble after nibble. As I wai ted for him to return again and again, I tried to imagine what he saw on arrival at the nest. Several gaping chick mouths into which he must constantly drop food to keep them alive. They must be ravenous and fast developing a taste for nibbles. When the grubs and flies have stirred he suspends the nibble run but until then, in the early morning LB's anxiety is palpable. One time he landed on the curvy vine with some flies in his beak, their legs sticking out at the side of his beak and he carefully fitted in a nibble as well. With this food combo he must know about balanced diets for the chicks instinctively. As he darts off in the same direction each time, I can only guess where the nest might be and wish I had a high tech camera positioned on it like they do in Spring Watch.

In the days that followed, LB came back and forth for food from sunrise to sunset. In the early morning when the chicks are at their hungriest, LB takes the nibbles and is so eager to deliver them, he flies in a direct line towards the secret nest and then minutes later returns for another refill. He has no time for cautionary zig-zagging to evade a possible predator on early lookout for any give away behaviour that could lead him to an easy meal. On the fourth or fifth mission LB swallows one for himself & then speeds off with one in his beak. I know that once he has swallowed a nibble for himself there will be a break in his delivery routine. He feeds the nestlings and his mate first then himself and then has a break before the next round of visits to the nooks & crannies that yield live food.

LB became more and more frantic for food. I saw him dart ing to all parts of the garden searching for the micro meals that his hatchlings needed. As his panic grew, I thought LB must be a first time parent grasping that the survival of his offspring was down to him, so his food deliveries needed to be unrelenting. For the first few days of their lives he would regurgitate tiny spiders and flies into the mouths of the weak and helpless hatchlings. I watched him flit from stacked plant pots to a pile of old bricks, from a tree mound to a newly watered patch of soil round a rose. He would appear on the curvy vine with wriggling food, a leggy spider and a squashed fly tight in his beak into which he also crammed a nibble.

Sometimes, as he opened up his already full beak to cram in the nibble, his live catch dropped out and the number of nibbles he took at one time made me think he pushed the nibbles straight into the gaping mouths of the hatchlings. He was constantly in frantic foraging mode. His role in his short life was to secure the next generation. I wondered how he sustained his stamina as he sped back to where the hatchlings were desperate for his beak-full.

Early one morning I noticed that LB had changed the flight path of his food deliveries. He now took the nibbles to the hedge left of the hut. This change in direction made me wonder if LB was supporting two nests. Research tells us that broods can overlap with the male robin feeding the chicks of one clutch while the female sits on the eggs of the next. Then again, LB could be using cunning tactics and the change in direction was intended to mis lead a watching predator. I have seen magpies high in the branches scanning for victims, swooping down on the birds pecking about on the ground under the feeder for seeds dislodged by the parrots, as they clumsily swung upside down on the wire frame, trying to push their hooked beaks through the bars. But LB seems to know how to keep below the radar. I am always aware how vulnerable litt le birds are, particularly when I am face to face with LB & see his spindly legs no wider than embroidery thread and his slender beak, his multi-functioning, life support ing tool little more than a centimetre long.

LB's nest life is secret. Of course I want to know how the nest lings are thriving but secrecy is very important to LB. If I went searching among the hedgerow branches and crumb ling walls, I would open up such an obvious trail for cats and other predators to follow. Getting too close to LB's sec ret sanctuary could alienate me for good and I can't risk violating the trust that exists between us. I can't be seen as a threat and I can't be more than a reliable and faithful food provider for LB, that is our pact. I am the one he searches for through the balcony window early in the morning when hunger and panic grip him. The nestlings will have twitched all night and now as light dawns they will be demanding to be fed.

LB continued to dart around catching live food and some
times landed on the curvy vine with a tiny worm dangling
from his beak to show me that he was still proficient at
collecting food from the garden. Then he crams in a nibble
to go with the worm, a fully balanced meal for a lucky
chick. His mate will still need to be fed & remembering
her drama queen act when he brought her in the evening
sunlight to visit the vine branches during their courtship,
I think she will be hard to please. Broods can overlap and
if she is preparing to lay another batch of eggs, she won't
have time to be capricious and will leave the survival of
the first hatchlings to LB. I want him to feel the satisfa
ction of having raised his family and seeing his chicks
grow up to be healthy independent robins. Maybe he will
guide the fledglings to my balcony and preserve the bond.

LB landed on the curvy vine early one morning with a dish
evelled patch of brown feathers on his chest where red ones
should be. Had he been attacked or been in a fight? I ponde
red the chances of such a small bird being able to defend hi
mself against the bigger garden residents should they turn

on him. The magpies would fight to the death and the cats would sport with his fragile little body. Suddenly, as he bent down to take another nibble from my fingers he went into attack mode, cocking up his tail feathers and singing a continuous high pitched warning song. Robins have two larynxes so can cope with food swallowing and singing at the same time. Then I saw another robin on the end of the balcony in the same fierce mode of attack. There was an ongoing territory dispute and a crisis was brewing. An early morning duel could explain why LB had a brown moth eaten patch on his breast. There could have been a fight and he had suffered peck attack. But he was still flying frantically back to his nest with nibbles. I must make sure he takes as many as he can, they will nourish him back to full plumage.

When I go out, I leave a small cluster of nibbles in the same spot on a table near the curvy vine for LB to find. He knows they are for him and exactly where to look. He can choose the time he wants to fly in for his snacks and I have watched him land on the table and bend down to select one from the pile. He takes one at a time so by watching the diminishing number of nibbles I know he is around and getting on with his life. But the nibble cluster is often stolen by squirrel bandits so I prefer to feed him directly to ensure that he gets them. He likes to have them handed to him particularly if he already has live food in his beak as there is less chance of losing it if I hold the nibble up to his beak. I don't know if LB connects the appearance of the self-service nibbles on the table with those that I hold up to him as I am there in person for the latter act but not the former. I like to think that offering him food with my fingers cements the trust between us and gives him a regular food routine that he can rely on. I believe I have been there at a crisis time in LB's life, spring was slow to arrive, he had a brood to feed and the intelligent little bird knew I would always be there to welcome him with what he needed most, food.

The garden has to provide food for many birds and the topped up bird feeder should improve their chances. The bigger predatory magpies, crows and jays want prey and are gifted with guile and intelligence and keen to exploit any situation. They scan around for small birds to watch and learn their routines, to see if they fly to and fro repeatedly in the same direction and so work out how they can get a small bird ready meal. They could track LB back to his nest to discover the chicks and they could spot me trying to get LB's attent ion and then track him back to his secret sanctuary. This heart breaking possibility makes me scan the tall trees for these cruel spies before I call LB over for a snack or put out a nibble cluster on the table. But LB is cautious and observant and seems to know instinctively to look around before dropping down onto the curvy vine.

I know LB often watches me when I go down the garden and if we have spotted one another, I call to him as I walk around. If we are close enough to have eye contact he glances away to look around for any signs of hostility. He could choose to ignore me, he has territory to protect and other robins wat ching and vying for garden space might deem him an out cast for fraternizing with a creature other than another robin and pick a fight. I do a quick sweep of the trees chec king for predators and go along with LB's cautious approach and look away, walk on a little and call to him again. If he wants to follow me and he usually does, he flies into the yew tree thick with branches which give him cover and viewing advantage to see when it is safe to fly down onto the curvy vine. Sometimes, his route to food is not via the yew tree but hopping from cane to cane and then onto the sloping balcony balustrade with legs astride in a comical pose, waiting for me to look up at him as I mount the steps up to the curvy vine where he knows he is safe and will be fed.

LB needs to respond to changes in the weather and adjust the timing of his tasks. I don't see him in sweltering heat, whirling winds or pelting rains. He seems to know when a torrential downpour is about to occur and may take temporary shelter but for longer bouts of bad weather and for darkness he will rest in his secret sanctuary. This location remains a mystery but I know the direction he flies to and fro regularly and his flight time takes only seconds so it is closeby. To go searching for it would damage the trust between us and discovery could prompt instant vacation, relocation or abandonment. LB and his mate will have built the cup nest out of dead leaves, moss and other thin strands of flotsam and lined it with hairs down. It will be hidden in shade to protect it from predators but will let in a few sun rays each day for the chicks who need it to manufacture vitamin D-3 to help them grow. LB cannot brood the eggs and newly hatched chicks because he does not have a brood patch, only his mate can do that. Food is not stored in a bird larder so he must forage during the daylight hours to keep his mate nourished so she has the energy and heat needed to incubate the brood. If he perished, his mate would continue to incubate but would not be able to sustain a feeding rota & would have to abandon the nest and all would be lost.

LB's routine changed to fit in with the brood cycle and his tasks were now linked to how his mate was coping with the egg laying and hatching. If she doesn't get enough nourishment she will not be able to produce her glossy blue eggs and if it is cold she may have to delay her egg laying because she can't generate enough heat to incubate them. She needs to be in good health before she is ready to lay the eggs and has to ovulate before egg making can begin, so if spring is late she won't be ready even though the nest is built. LB has to be sensitive to all these variables. The pair have to know about the humidity and the temperature around the nest and will have to make sure that no part of a tiny growing chick gets dried out or stuck to the shell. LB has so many tasks, some days when he lands on the curvy vine for food he looks tired and weary. He will spend every waking hour from sunrise to sunset searching for food, probing the garden for tiny insects, spiders and other squashy creepy crawlies. I imagine him landing on the edge of the nest facing newly hatched, blind and helpless little chicks, with their mouths wide open to receive his regurgitated mouthfuls. Their microbe smoothie contains nibble so will bolster nutrition. Keeping the chicks alive is LB's responsibility, he will need all his ingenuity to succeed.

I think LB and his mate are on their second brood which seems to have overlapped the first. While his mate is concentrating on incubating the second brood, LB will have to supervise the first fledglings and sure enough one morning towards the midde of June I saw LB with them. This was the first time I had seen his offspring as fledged baby birds. Two new fledglings were clinging to him and hugging the ground beneath the yew tree as he was collecting titbits from the soil and flying over to collect nibbles from me to feed into their beaks. A little crowd of baby birds gathered around him and it became hard to know whether all the birds LB was minding were his. Paternal robins can gather up a crèche of other fledglings as well as their own.

The little fledglings had speckled feathers which marked them out as juveniles. They won't have much time to learn how to protect themselves, LB would have to teach them survival skills in a very short space of time. Soon he would have to leave them to fend for themselves while he concentrated on the next brood. Adult robins won't attack juveniles without red breasts, nature's way of protecting the young for as long as possible. Their youth and inexperience made them very vulnerable to predators but not to other robins. Being taken as serious contenders in territorial disputes would come when they acquired red breast feathers during their first moult at about three months old. As LB took the nibbles I held out for him, I wondered if one of the juveniles would watch and learn and one day, venture to appear on the curvy vine. He continued his regular rounds of garden

foraging, darting and jumping from spot to spot, feed
ing micro meals to the fledglings and towards the end of
the day, he came to the curvy vine for the last time to
swallow a couple of nibbles and take one away in his
beak, presumably for his mate. As he perched, I saw a
distinctive brown feather patch on his breast and wond
ered if this was the beginning of his summer moult.

Not long after the crèche gathering, there seemed to be a sad
ness about the way LB came to the curvy vine for food which
made me wonder if he had lost some of his young and was
grieving for them. From the constant care he gave the fledg
lings, I could see that he was aware of the hidden dangers
that he had to shield them from. The cats were extra creepy
they knew there was sport to be had by stalking the ground
for little bird bodies who were hopping and peering around,
waiting to be ushered by their parent and for food to be put
into their beaks. During one of my vigilante stakeouts
trying to protect LB, I saw a particularly lethal cat creep
under the lily leaves to lie in wait for a hapless fledgling.
I saw that LB had also spotted his slinky body disappear

under the leaves and quickly guide the fledgling he had
with him up into the yew tree. LB's quick thinking had av
erted a potential crisis and there was no need to rescue a
damaged little bird this time. The cat received a nasty shock
which would keep it out of my garden for quite awhile.

The anxiety from a potential fledgling rescue returned when
a disabled neighbour from an adjacent garden made the diff
icult journey to my door holding a carrier bag containing
a baby bird for me to inspect. Within the folds of the bag was
a damaged robin rescued from the mauling paws of a cat.
My heart started to beat very fast, was LB inside the carrier
bag? I dropped to the floor holding the bag and rolled back
the folds very carefully. Would I find LB damaged or even
dead? I peeled back the final fold and gently cupped my
hand over the little robin who lay motionless. As I held
him I could feel his heart racing and thought he could ex
plode out of fear. I saw that his little breast feathers were
a dull browny red so it wasn't LB. I felt such relief but I
still had a young bird to rescue and needed to see what
damage there was. He remained still as my hand around

him prevented any flapping that could hurt him & I gently
removed a tiny feather that was stuck to one of his eyes. I
couldn't see any blood or obvious puncture wounds so thought
he could rest in a recovery box. I started to release the pressure
of my hand that cupped him and with a quick little flutter,
he flew up to land on the top of a wall picture disturbing
copious amounts of dust as his feet settled on the frame. So
his wings weren't damaged but was he one of LB's little brood.
I began to talk to him in the same way I talked to LB and he
started to take notice and stare at me. Then after a few more
minutes recovery, he flew up to a book shelf and landed on a
tall book and peered down into the room and at me as I cont
ued to talk to him in what I hoped was a reassuring way.

Seeing that he was still and not panicking, I went slowly to
open the balcony door to let the garden sounds in, perhaps
there was a frantic parent bird calling to find their missing
offspring. The little robin on the tall book perked up and
listened to the sounds. It was at this point that LB sudden
ly appeared and hovered in his humming bird posture mid-
air to peer in through the open doorway into the room, a

trick he had performed before when he wanted my attention
for food. He must have seen the commotion and come over
to investigate. What an extraordiny coincidence LB chose
that moment to hover in the very doorway the baby robin
was watching from his perch on the book shelf.

As LB hovered, flapping his wings very fast to steady himself
waiting for me to offer him the nibble bowl, the recuperating
little bird was watching this very unusual behaviour. Would
he recognise LB as a relative, but as LB arranged two nibbles

into his beak there was no recognition and he quickly flew off.

As baby bird watched LB fly off into the sunshine he came to life and suddenly launched himself off the book & flew over to the curtain rail next to the door opening out onto the garden. As he flew across the room, some of his soft downy feathers floated down so there must have been an attack on him dislodging some of his baby plumage and I could see that his tail feathers were crooked. He flew onto the top of the door and looked out into the garden, listening to the cheepings and I saw there was some recognition. He was more confident, turned his head to one side and listened to the bird calls. Suddenly, he bounced up and flew out of the doorway into the sunshine. The crisis was over and baby bird was free again to join his family. I was very relieved but sad too as the episode could have ended so differently. I decided to increase my vigilante surveillance of my garden to protect LB as he fed and taught his little brood about the perils that lurked in the garden. LB was responsible for showing them how to avoid danger and be independent. Cats and other predators beware.

The day I saw LB have a bath convinced me he had a playful
side. One of my first encounters with him had been with my
garden hose. I was using it again to water the plants & this
time he wanted to perch on it close to the nozzle as it delivered
its spray. I reached the plants by the bird bath, an old BBQ tray
I had adapted with flat stones in the centre where small
birds could drink and wade into the shallow water pool.

As I sprayed plants inches away from my avian paddling pool,
LB hopped from the hose on to a stone in the middle of it and
then into the water tossing the spray around his feathers and
diving his head to make more spray. We watched one another,
it was as if he wanted the display to be enjoyed by me as well.

He splashed around for several minutes, a break from end less rounds of brood-feeding. Getting himself wet revealed more brown feathers on his breast, his summer moult had begun and the patches made him all the more endearing.

One sunny June morning, three robins were engaged in a dra matic display around the base of the bird-feeder. Surely it couldn't be food that was making them so excitable. Three meant one could be an intruder. LB's first set of fledglings were launched so his search for food was now less frantic. The robins were displaying their cocked tail feathers and div ing at one another. I could see one of them was LB from the distinctive brown moult patch on his chest. It was possible that one of the robins was his mate taking a break from second brood nest preparation or a new available female. The jumping and strutting was escalating so there had to be two males. All three robins had red breasts so were mature adults and would be expected to know how to fight for superiority. Disputes can start with males sing ing at each other and flying into high perches to show off their red breasts in a more threatening stance. If one robin doesn't defer to the other, resulting fights can lead to injury and even death. Thankfully, this quarrel died down and eventually LB left the fray & wisely flew off towards his secret hideaway avoiding any damage.

One hot sultry day in July, LB failed to show up. His garden activities were much less busy at this time and there were no fledglings or his mate in attendance. The day went by and still he hadn't taken the couple of nibbles I had left for him on the table. I decided to sit out and call him to let him know I was there. Day became a balmy evening, the tits were busy finding midges high up in the yew branches and the ground birds were hovering underneath the bird feeder for seeds to drop, dislodged by greedy parrots as they pushed their large heads through bars not designed for birds of their size. This seed sprinkle was a source of bounty for several creatures & I had seen LB peck up some in the past. But still he did not appear and it was late. He is up with the dawn but when the light fades and night time creatures like owls and foxes start to stir, he knows to stay hidden. In all the weeks I had known LB, I would see him at least once in a day. I knew one day LB would be worn out and unable to continue his life and his routines. But surely not yet, not now.

I started to think of plausible reasons for his absence. The worst and ever present fear was predatory harm, but LB was smart and quick, he always paused before making a move and would take avoidance action if there was danger close by. But his nest was more vulnerable and his family might have been discovered and destroyed. LB seemed to be a loner again, he hadn't sung for quite awhile and he ate only for himself and didn't take a nibble away with him to test tap on a branch as he had done in his courtship days. Then again his mate might be sitting tight on the nest creating another brood and LB needed to be there to protect her. I wanted to attribute his absence to normal behaviour in the robin cycle. I couldn't think of the garden without him.

If LB's summer moult was in full swing he would be on his own again and need to keep a low profile. He would know that he didn't look right and be vulnerable to attack if he was in-full view of other birds. Daily feather loss would make him feel itchy and irritable as new feather stubs poked through. If this was his first moult he wouldn't know what to expect, he might think that his feathers had fallen out permanently. Losing his insulation feathers would make him susceptible to the cold at night, he needed to hide away in the hedgerow foliage and wait for his new feathers to grow. I was relieved that none of the distressing reasons that I had considered for his absence were valid on this occasion. I had to be patient, LB would appear when he had the strength to fly.

I spotted him looking small and fragile underneath the bird feeder hopping among the flotsam. I called and he looked up but would he come over? LB could be capricious and he was vulnerable now, finding a morsel on the ground would be compelling. He stared at me and then bounced up to fly onto a nearby cane and then onto the curvy vine to present him self for food. He looked dishevelled and scruffy, patches of brown feathers were showing through the red breast ones. I gently offered nibbles to him and he bent his head to take and swallow a couple before flying onto a yew branch where he paused to bend his head under his wings and give some tugs and preens to his feathers. He must be in full moult now and would have to endure the loss of the old feathers and growth of the new ones. He will have to make ongoing adjustments to his flight control and balance. This could take weeks and in the meantime he would need mild weather, cold nights & torrential rain could harm his denuded body.

During the next few weeks LB would need energy & endurance to get through his moult. He must remain inconspicuous & hidden among protective fronds and avoid the attention of garden spies and sneaky prowling feline hunters. He had worked so hard to raise his broods, now his feathers were worn out and he was alone again. Moulting was a solitary affair, his mate would have left him to undergo her own plumage transformation. New strong feathers were needed to protect against the winter weather, and as they grew through the skin LB would need time to adjust and recuperate. He hadn't sung for a while, the garden was quiet, all the adult robins would be moulting and keeping out of sight. His visits to me were less frequent, he didn't take many nibbles and there

was a summer garden food fest there for the finding. I had to
call LB several times now before he came. Tree branches heavy
with leaves would absorb sound and make it harder for LB to
hear me but I knew his hearing was good, I had seen him cock
his head to one side as he listened to bird song. Sometimes, he
would suddenly appear on the curvy vine without being called &
look up for me and then I felt so relieved he still trusted me.

It would take some time for new breast plumage to grow again
and cover the patches made by moulting feathers. His appetite
was returning and when I opened the curtains in the morning

I didn't have to wait long before he flew onto the curvy vine for food. The leaves around his bit of the vine had grown bigger than he was and during one visit they frightened him as they brushed against his wings in the wind so I pruned them. Now he could land freely again and stare at me as I held up a nibble to his beak before bending his head to grasp and then hold it in his beak before swallowing it. I made the nibbles smaller so he would need more & stay longer until he was full. Now he took three or four, pausing to swallow each of them before flying onto a yew branch to wipe his beak.

LB kept a low profile for the next few weeks and when I saw him briefly he seemed forlorn. Possibly he was pining over the departure of his fledglings and his mate. I wanted to believe they were still a duo but I knew they didn't pair for life and she would need to hide away for her own moult. Perhaps after successfully brooding the babies she had abandoned LB in favour of independent territory and a new partner. She had been nest bound at the beginning of their brooding so wouldn't have seen him frantically flying like a torpedo through the trees to my balcony time and time again for nibbles to carry back to the hungry fledglings. This level of devotion deserved to be rewarded with constancy. But no, LB was single again and enduring his moult. I would have to wait until next year to see whether they paired up again.

By mid-August LB was in full moult and looking even more forlorn. On one occasion when coming for food he stumbled while landing on the edge of a yew branch prior to flying down onto the curvy vine, a manoeuvre he had done many times. He was usually very nimble but the loss of a large number of tail feathers was interfering with his ballast. Not long after this visit he returned stumbling again as he landed without any tail feathers at all. He looked so small and vulnerable with a few white feathers sticking out where his tail should be and he was gripping the vine tight with his spindly little feet to keep his balance. LB must have summoned a lot of courage to fly over having lost such a substantial amount of plumage. If this was his first moult as I suspected, LB would not know how long he had to wait for the new feathers to grow, he might think his feathers were gone for ever, a victim of feather blight.

Later I saw LB in the garden perching motionless on a brick pondering on his next move. He saw another robin who hadn't lost any tail feathers which I imagine made him feel worse and he flew off avoiding any confrontation.

While LB was growing new feathers he was shy and unsure of himself. His wing feathers were dropping out so daily flight adjustments would have to be made to ensure that he could still function. He would be a fledgling again not knowing if he could rely on full flight control when he took off. When he did come for food he perched on a high branch in the yew tree looking down at me pondering for some time before dropping down to the curvy vine. He was nervous and knew his vulnerability would make him an easy target but none of his temporary frailty prevented his sharp eyes scanning the panorama whenever he was waiting to fly down for food. Usually, he would face me when he landed on the curvy vine but if he had spotted a noise or movement that was obtrusive and out of place, he turned round to face into the garden looking up and around to make sure that he was safe.

Each day, more moult patches and new tufts of bedraggled feathers growing at random angles made LB look tousled and dishevelled. Now when he landed on the curvy vine, he was making brave efforts to avoid wobbling. Without tail feathers to help him steady himself he looked small and slight. He looked at me when taking the nibbles as if he was ashamed of the way he looked, his ebullient bounce was gone and his confidence seemed at a low ebb. Usually robins in full moult hide out of sight for the duration, their defences are compromised and they are exposed to predators. I felt humbled that LB felt safe enough to risk coming for nibbles. I hoped their high energy nourishment would help him endure his moult and grow strong tail feathers. His hideaway had kept him safe hitherto, LB needed to bide his time until his winter plumage had grown. Increased visits for food meant he was more active and getting stronger.

When LB didn't appear I went round the garden calling him. It was possible that lush summer hedgerow growth muffled his hearing. On one occasion after several calls he appeared with his feathers sodden and dishevelled. He had probably been in the bird bath to ease the feather itching. Little tufts were sticking out of his head and new brown feathers growing out unevenly on his chest made him look shabby. Loss of feathers was altering his shape so that his appearance was constantly changing. Sometimes he was more red than brown, sometimes more brown than red, and white feathers round his middle stuck out at odd angles. He didn't have enough feathers to feel complete & his demeanour was sad.

LB had led a solitary life since his moult had started. He kept out of sight and I only saw him once a day on the curvy vine. I would usually have to call several times before he appeared, he didn't stay long and after taking a couple of nibbles was eager to fly back to the hedgerow. Day by day his plumage started to look less scruffy and his outline a little more substantial. Then one day in the late morning sunshine I saw him hopping and diving for tiny creatures under the bird feeder. After several minutes he flew up onto the rim of the bird bath and watched ahead in the stillness for a long time. After pondering for several more minutes he flew down onto a stone in the middle of the bath for a drink, taking

time to swallow in between sips. when he was full he flew up
onto a moss covered brick on top of a tall plant pot. This
was another of his favourite ponder spots where he could
perch, the moss making a soft surface for his feet to grip.

LB likes to perch and ponder in the sunshine and then drop
down to the ground to catch flies and other micro creatures,
making darting movements and energetic ground tugs to
pull up worms. As evening approached, I usually watered
the garden and very often I saw him on the canes & branch
es watching my progress. He had seen me do this many

times and knew that I put the equipment away on the balcony when the job was done. Anticipating that I would arrive at the balcony fence, he sat on it so that he would be there to catch my eye. LB is nervous and needs to feel it is safe so I perform my sing-song greeting as I approach. When I get close up to him I bend my head down to be on a level with his and hold a nibble up to him. He bends his head to take it and swallows another three but he is still not the effervescent little bird he was when tending his family.

Then a remarkable thing happened. I was watering the garden with the hose while wearing my red jacket which I wore for garden work. I thought it was a colour LB liked and made me look robin-friendly. I had been spraying water on the plants for about half an hour unaware that I was being watched, when I heard a distinctive short sharp wheeze of a bird song. I knew it was a robin so instinctively looked up to my balcony fence and sure enough LB was perching on it staring at me and calling to let me know he was there. As soon as I looked up he called again with more high pitched wheezes. It was not random it was intentional. LB had seen me from a distance in the garden, flown over to his usual balcony spot and was trying to get my attention so I would feed him. I put the hose down and walked towards the steps that I needed to climb to get to where he was perching.

LB was still calling in short sharp wheezes and watching me as I approached the sloping balustrade on which he was now standing so that his legs were splayed in a comical pose. I had started to talk to him so that he wouldn't lose his confidence as he waited for me to get close before hopping over to the usual nibble handover spot. I picked up the nibble bowl and sat down beside him so that our heads were on the same level and held up a nibble to him. He swallowed it making a short burst of wheeze song in his throat and waited to be handed another which he took repeating the same song burst, which by now I was convinced was a sort of greet and thankyou utterance. He took a third and with another quick wheeze flew up onto a yew branch to wipe his beak from side to side. This was the first time LB had called me directly. He had seen me but deduced I hadn't seen him and he didn't want to wait any longer

so he started his short sharp wheezy call to get my attention.
which had worked and as he watched me look up towards
him he sustained his call to ensure that I would come and
feed him. This felt momentous, a robin had learned to
summon a person from a distance, a real vice-versa
situation. I could call LB and he could call me.

Now that LB knows I can hear and respond to his call, he used
this clever tactic on another occasion to get my attention.
He was concealed in the yew tree which provides good cover
and is usually his first approach before hopping down onto
the curvy vine. He was making his wheezy call so I looked
up to the yew tree and called to him. A branch wobbled so I
knew he was making a move. He hopped out of the dense
foliage and down onto the curvy vine still singing wheezily
and looked up ready to receive food. It was four nibbles now &
he swallowed them in quick succession. His new feathers
were growing and his appetite was returning. Prolonged
wet weather in the garden prevented flies and spiders
showing themselves and I imagined that a very bad
summer could be very serious for robins who needed to

lie low during their summer moult and rely on easily accessible food to sustain them. LB sang wheezily in bet ween swallowing as if he was on a musical roll.

Not long after, LB performed another smart move. He was waiting in the yew tree outside my window hoping that I would appear but I was nowhere to be seen as I was inside working and unaware of him. The balcony door was shut but the curtains were open so he knew I was in somewhere. He must have waited and waited until his hunger forced him to try to get my attention in a very unusual way. I heard the brush of movement against the window and knew it was the sound of a bird brushing against glass. I went to open the balcony door and sure enough LB was standing on the curvy vine looking up at the doorway. He had done the only thing he could think of to make me come out, he would try to get in through the glass to alert me. What ingenuity, I was full of praise in my sing song greeting as he swallowed several nibbles. He was still without tail feathers and his lower body looked like a small round ball of down. He would still feel very vulnerable at this stage in his moult so brushing his little body against the glass to alert me was very brave.

About this time in August, another robin was looking very at home in the garden, could he be one of LB's rivals? Usually, LB flew in from the hedge but this robin still had tail fea thers and inhabited the trees and ground around the bird feeder. His moult was not as advanced as LB's so he still looked bold and commanding. I tried calling him but my overtures were ignored, this robin was not going to get personal. Other robins around the garden would be in various stages of moult but it didn't stop them singing their chorus of songs high in the branches early on sunny mornings announcing their territorial claims. This perfo rmance was very competitive and I saw two robins fly off in pursuit of one another reminding me that a serious clash could easily disrupt the choral harmony. Territorial disputes start with male robins singing at each other from a perch high up in the trees where they compete to show off their red breasts. LB needed to stay out of this fray as he lacked enough body feathers to bulk up his size and he did not have the full armour of flight feathers yet. By the time autumn became winter, robins would be with adult plumage ready to sing their late autumn songs, subdued and melancholy. Months of icy cold weather had to be endured before spring appeared again when robin songs would become more upbeat, declaring interest in future pairings.

Not long after, I was in the garden inspecting some bulbs which were stored in a pot with a brick on top to keep the squirrels out. LB was close by watching what I was doing from the safety of a nearby branch. Any movement amongst garden paraphernalia excited him because he knew it could reveal tiny creatures for a meal. When I picked up the brick I saw that a very energetic centipede was running round it. I knew LB liked centipedes so I held it up to him hoping he would see it and grab it before it escaped. But the speeding centipede was running away so fast I had to keep turning the brick round to prevent it from dropping off. I could see that LB was confused so I talked soothingly to him as I held up the rotating brick. Would he see the racing centipede or would the brick appear as a large menacing shape?

I wanted him to connect the held up brick with the same food offer gesture I used when I held up nibbles to him on the vine, a position LB was used to and where he felt safe. I was hoping he would trust me enough to realise I was offering him so me food even though it looked very different to a nibble & was in an unfamiliar location. But LB was becoming more and more agitated by the brick and couldn't stay long enough to spot the centipede so he flew back into the foli age for safety. His innate sense of self-preservation had dictated retreat. On reflection, this was a big relief. LB was not a bird to take risks so I could be more assured that when I didn't know where he was which was most of the time, he was most likely somewhere safe.

During these transition days when old feathers were dropping out and new ones were poking through, LB was hungry and took several nibbles in quick succession swallowing them fast, flying off with the last one in his beak. This routine conti nued for many days which were getting shorter and were suffused with bright autumn light. Once the sunlight beg an to fade, dusk came quickly and LB never ventured out

in the half-light to the curvy vine for food. Dim light brought
compromised visibility at a time when predacious nocturnal
creatures were waking up. LB had to judge when it was safe
to make the last visit of the day to fill up with food for
the night. There were many dark hours to wait before it
was light enough to return for his first meal of the day.

Day by day LB's feathers grew stronger around the contours
of his little body. When he landed on the vine he made his
chest feathers swell in a wavy ripple. The white feathers
looked as though they had been freshly laundered & the
red ones were a flame colour and shone bright when he
was in the garden so I could see him. Then one day he
landed on the fence next to me and began to sing a loud
and joyful song as if he was pleased to show off his new
apparel. He took time to bend his head for each nibble &
as he turned round to fly away with one in his beak I saw
his sturdy tail feathers. LB was becoming himself again.

When I haven't seen LB for a time & wear different coloured

clothes he is cautious and unsure about flying down to the curvy vine. Although I have tamed him to respond to my call and reward him with food which is what he wants most of all, I have to remember that he is a wild bird. He doesn't respond to strangers, being suspicious about the movements of all creatures including humans makes good survival sense. He must be alert to threat of any kind. Some times he is very jumpy when coming to the curvy vine, any sudden movement can make him leap high into the air. But LB usually recovers and calms down when I talk to him. I have watched him looking at me from a distance when I call, pondering on whether to fly over or not. His instinctive caution forces him to make further safety checks. Sometimes on approach he lands in amongst the vine leaves which shield him so he can peep through the gaps to see if it is safe to proceed. He sees me looking at him and he stares back playing peek-a-boo. I reassure him with more singsong words before he sum mons up enough courage to jump up and fly over. Once he is close enough for me to feed him, he is calm and watches my fingers hold up the nibbles which he swallows quickly, carrying the last one in his beak to tap on a yew branch before swallowing it and flying off to another routine stop.

The season was changing fast now, the days and nights were becoming much colder. Nature was keeping pace with a seasonably predictable timetable. The plentiful creepy crawlies that LB had foraged for throughout the summer were fast disappearing and he would need new routines for winter. His moult over, his new-feathers would help him to survive the bitter night air as it swirled around his night-time hedgerow hideaway. LB had started to sing again and frequently sang bursts of his reedy warble song to me as he landed on the fence waiting for food. Watching LB sing this unique warbly song only inches away as he looked at me will be one of my abiding memories of him.

The lengthening of the nights delayed the dawn so the robins had to adjust the start of their morning melodies. Their cue was the rising autumn sun spreading a veneer of golden light among the tall trees where the robin songsters sought high branches to start their song cycles. Then they unleashed high trills and low shrills and all the warble descants in between. Now that LB was bedecked in full plumage he would feel the pull of this musical convention and his morning priority was now singing songs not food. He wanted his part in the morning chorus. I watched him fly up high into the branches to take his place and he only came for food later when the songs had been sung.

Now, on a fresh sunny morning when I stepped out into the garden, I could hear the songbird symphony and the more I listened the more distinctively unique it was. I marvelled at LB singing his part with the other songsters proclaiming

their jubilation and luxuriating in the expanse of bright
clear blue sky and endless panoramas from their treetop
domain. Our own choral music has its origins in the latin
Gregorian chants sung centuries ago in the monasteries.
There is a record as early as the 6th century of a robin
taking food from a human. So I will call this performance
the Erithacus Rubecula Chorus, The Robin Chorus. Know
ing that these sweet songs are the same seasonal songs
heard by our ancestors from centuries past is comforting.
I stared up into the branches but couldn't see the music mak
ers, they were invisible to me but not to each other. Every so
often, a member of the chorus selected a new perch. I can
only guess what secret communications were being establ
ished as the notes flowed forth, bird dalliances and displays
for future pairings perhaps or communal rejoicing that
their time was coming round again. The naked trees of win
ter and the snow carpets would provide their own backdrop.

when I am very close to LB on the curvy vine I can see the full
transformation, his new plumage is clean and pristine and
his outline is sleek. The bright vermillion red breast feathers

are in perfect layers round the top part of his body and their surface is smooth and glossy. New, strong olive brown tail and wing feathers have given him a sturdy frame and he no longer wobbles as he hops onto the end of the branch to survey the scene before dropping down onto the curvy vine. He swells out his red breast with a ripple of pleasure and pride. Perhaps he is showing me what a handsome bird he is now that he can show off the all new feathers look. His mood is buoyant and he is still as hungry as he ever was. He has started to sing a few cheepy notes again in between taking the nibbles particularly in the early morning which is his favourite time of the day. These notes were just a prelude of what was to come.

Having passed the equinox the morning chill was a perman ent feature of the dawn and had depleted the availability of live food for LB to find. He was in a hurry when he came for food now and when I wasn't quick enough handing him the nibbles he hopped into the nibble dish and helped him self, eager to be off on his mission which was to sing. As winter winds wailed through the empty trees with their bare black branches silhouetted against the sky, LB flew up into them to sing for hours at a time. He abandoned food visits and when he missed a day after a torrential night storm, I started to think he was damaged or had perished.

And then I saw him in full throated song as if compelled to fulfil ancestral callings to pervade the atmosphere with shrill harmonics and passionate cadences. His throaty cheepings had developed into a choral masterpiece which he was joyously performing. I could see that his whole body throbbed with the vocal vibrations as he gave his recital and he was not alone, other robin melodies were all around joining in the chorus. These songs held musical messages understood only by the singers. LB was high among the branches in full throttle reminding his clan that he was entering the robin arena and showing that he was worthy of taking his place among the feathered choristers. He would be hoping that an enraptured listener would eventually fall for his musical charms and want to join him in building a new nest home. This was the message and the meaning.

Winter had arrived and the seasons had turned full circle. Now LB's priority in the first light of the morning was making music not searching for food. He wanted to soar high up into the branches to open his throat and deliver his ancestral refrains. LB had reached a milestone. He had survived a busy year and was now preparing to endure the next and whatever adventures that would bring. His renewed energy and new winter plumage had equipped him for another chapter in his life and his unique early morning song I can hear him sing is announcing that he is ready to live it. But events unfolding in this next chapter would prove to be as elusive as LB. I would have to face up to the reality that LB had already lived his final chapter.

The last time I saw LB forage in the garden, he was toss
ing up leaves bigger than himself and lobbing them
with gusto over his head to uncover the hidden food.
He made a fleeting visit to the curvy vine & as he took
a nibble he became distracted by a robin singing in the
tree above. He tilted his head still holding the nibble to
listen intently and then flew away. After that LB's visits
to the curvy vine stopped and he disappeared. As I contin
ued my daily calling in the hope that he would return,
I could hear loud, melodic robin songs piercing the cold
air in the tall trees. Back came an almost instantaneous
reply and I wanted to believe it was LB, not a coincidence
but his way of acknowledging my call and confirming
that he was alive. The songs were constant, a refrain would
elicit a reply, prompt more singing and more replies. This
song cycle continued all day and evening for weeks. But
I couldn't distinguish and be certain that one of the sing
ers aloft in the trees was LB. I wanted to believe that he was
part of the chorus but to locate robins singing high in the
trees is not easy. His visits had been so regular but now LB
had vanished. I had to face a darker explanation for
his disappearance, he had perished and I would never see

him on the curvy vine again. I had to absorb this distre
ssing reality and was filled with sadness when I looked
at the curvy vine stark in the winter cold.

And then one day while I was working in the garden, a
robin landed on a branch and began to sing a loud-full
throttled song close enough for me to see his beak open
to let out the joyous melodies. I stood looking up at him

and began my LB greeting song. He didn't fly away & sang
for over five minutes so vigorously that his body shook
with the effort. It must be LB, any other robin would have
flown away immediately I approached. I started to believe
that LB was one of the tree top songsters. Robins were now
in a new cycle, their priorities had changed and they were
now intent on proclaiming themselves ready to compete
for a partner. LB would be strutting his stuff to impress
a new mate. Robins were flying off two by two, the mild
winter weather could have deluded them into thinking
that it was almost spring and that there was an urgency
for them to pair up and start to build a nest.

One February morning just as I had finished my token
call to LB, suddenly out of the blue he landed on the vine
branches. My surprised movement made him jump in the
air as I quickly started my greeting song. He tentatively
edged closer and landed on the fence by my chair that I
sat down on so that my face was on a level with his. I
didn't think I would see LB again but here he was. He
looked at me, tilted his head upwards and started to sing.

Having lived on music for the last three months a pang
of hunger had brought him back to the curvy vine. The
cycle had changed. Fewer robins were flying into the trees.

LB was hungry and in nest mode so he must have found
a mate to share his life with. I thought it remarkable
that he remembered the nibbles and our feeding routine.
In the days that followed, he resumed his regular
morning visits and three days later he was already
waiting on the curvy vine for me to open the curtains.
His memory was sharp and he remembered detail. I was
now convinced that he had been watching & listening
to my calling him throughout the past months. His shy
mate came with him to wait in the yew branches as
he collected her a portion of nibbles to pop into her beak.
To vary his diet, I collected small worms which I kept
in a container. LB watched intently as I pulled off
the lid and offered him a wriggling delicacy to seize
and swallow. I must build up his strength for paternity
again, there will be eggs to nurture and soon new life
will break through. LB's year had begun again.